The INSIDE GUIDE

STEM CAREERS

# Jobs in Artificial Intelligence

By Beatrice Harris

Cavendish Square

Published in 2024 by Cavendish Square Publishing, LLC
2544 Clinton Street Buffalo, NY 14224



First Edition



Cataloging-in-Publication Data

Names: Harris, Beatrice.
Title: Jobs in artificial intelligence / Beatrice Harris.
Description: Buffalo, New York : Cavendish Square Publishing, 2024. | Series: Inside guide: STEM careers | Includes glossary and index.
Identifiers: ISBN 9781502670274 (pbk.) | ISBN 9781502670281 (library bound) | ISBN 9781502670298 (ebook)
Subjects: LCSH:
Artificial intelligence–Vocational guidance–Juvenile literature. | Artificial intelligence–Juvenile literature.
Classification: LCC Q335.4 H36 2024 | DDC 006.3023-dc23

Editor: Therese Shea
Copyeditor: Shannon Harts
Designer: Deanna Lepovich

The photographs in this book are used by permission and through the courtesy of: Cover, pp. 15, 16, 18 Gorodenkoff/Shutterstock.com; p. 4 fornStudio/Shutterstock.com; p. 6 Colossus/Wikimedia Commons; p. 7 VineStar/Shutterstock.com; p. 8 Turing in 1935/Wikimedia Commons; p. 9 EF Stock/Shutterstock.com; p. 10 Kateryna Reka/Shutterstock.com; p. 12 ymphotos/Shutterstock.com; p. 13 TY Lim/Shutterstock.com; p. 14 r.classen/Shutterstock.com; p. 19 fizkes/Shutterstock.com; p. 20 PaO_STUDIO/Shuttersrtock.com; p. 21 REDPIXEL.PL/Shutterstock.com; p. 22 Suwin/Shutterstock.com; p. 24 Frame Stock Footage/Shutterstock.com; p. 25 (main) Pixel B/Shutterstock.com; p. 25 (inset) Austin Nooe/Shutterstock.com; p. 26 NASA/JPL-Caltech/MSSS; p. 27 Zapp2Photo/Shutterstock.com; p. 29 (left) Stefano Mazzola/Shutterstock.com; p. 29 (right) Milan Ilic Photographer/Shutterstock.com.



CPSIA compliance information: Batch #CWCSQ24: For further information contact Cavendish Square Publishing LLC at 1-877-980-4450.

Printed in the United States of America

# CONTENTS

Many of the devices we use today for communication and information are computers. Most of these devices use artificial intelligence.

Chapter One

# AI IS THE FUTURE

Have you ever thought about what a computer is? It's any device that works with information. In modern times, though, computers are thought of as electronic machines that can perform a task according to a set of instructions called a program. These tasks may include storing data, doing calculations, controlling another machine, or interacting with people in some way, as in a game.

Computers as we think of them have been around for less than a century. The first were huge machines, taking up entire rooms. Today's computers are much smaller—and much more powerful.

A branch of computer science is taking computers a step further, making them "smarter" and capable of performing more tasks than ever before. It's called artificial intelligence, or AI.

**Fast Fact**

**Many Black women did math calculations for early U.S. space missions. They were called human computers.**

## What Is Artificial Intelligence?

Intelligence is the ability to gain knowledge and use it in new ways. Humans are naturally intelligent. Our brains acquire facts and learn skills.

**This Colossus Mark 2 computer was created toward the end of World War II (1939–1945) to break secret codes.**

Artificial intelligence is **technology** that can use data to perform tasks or solve problems like a human being can. This kind of intelligence is called artificial because computers must be programmed—by a human—to accomplish this. AI isn't a physical computer part, but rather programming used by the computer's central processing unit (CPU), also called its microprocessor.

Earlier computers didn't have the processing speed or data storage to handle much programming. In the 1980s, advances in technology allowed AI to move forward. Since then, computers have been programmed to understand human language, play **complex** games,

**This computer is simulating human intelligence. That means it's imitating what a human would do. This is accomplished through programming.**

and control robotic parts, among many other tasks. AI is increasingly a part of our daily lives, whether we're aware of it or not.

**Fast Fact**

In 1952, computer scientist Arthur Samuel created a checkers-playing computer program, the first program to learn how to play a game.

## Cutting-Edge Careers

AI programming can perform as well as humans in a variety of ways and in a variety of places. It's even surpassed us in some tasks. However, it has not matched the human brain in its complexity ... yet!

## *ALAN TURING'S TEST*

Alan Turing (1912–1954) was one of the earliest scientists to work in the field of artificial intelligence. In 1935, Turing talked about the concept of a computerlike machine with the memory and capability to improve its own programming. His idea came true with modern computers.

In 1950, Alan Turing created what is now called the Turing test to determine if a computer can "think" like a human. Basically, a human questioning a computer and another human being should be unable to tell the difference between the computer's answers and the person's answers. Some think a program called Eugene Goostman passed the Turing test in 2014. It tricked 33 percent of the judges.

During World War II, Alan Turing worked with others to build a machine that translated coded German military information.

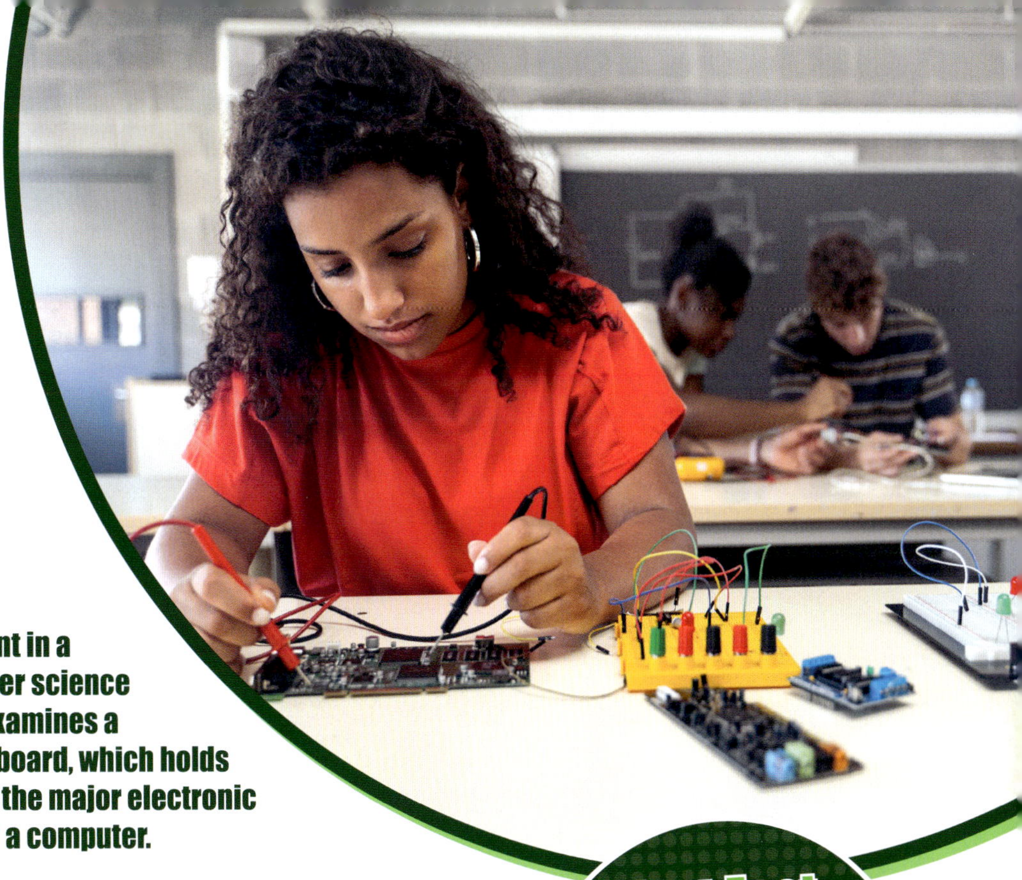

A student in a computer science class examines a motherboard, which holds most of the major electronic parts of a computer.

Many people are excited about the ways that smarter AI can help humankind in the years ahead. They're choosing careers to get involved in and further advance artificial intelligence. There isn't just one kind of AI job to pursue—or even just one career path. However, all require STEM (science, technology, engineering, and math) knowledge, skills, and experience. Most AI comes about because of the **collaboration** of many people. In this book, you'll learn about several jobs that create and support AI and some of the areas in which artificial intelligence is making a major impact.

**Fast Fact**

The Loebner Prize is a yearly contest that rewards AI computer programs considered by the judges to be the most humanlike.

Computer programming languages look very different from the languages humans use.

Chapter Two

# AI LEARNING AND LANGUAGE

Artificially intelligent computers get their intelligence from software, which is a collection of programs that work together. Software is written in special languages that the computer can understand. Python, C++, and Java are computer programming languages often used for AI programming. As computers have become more powerful, programmers have written increasingly more complex programs to grow the field of artificial intelligence. Some careers connected to AI deal with creating software to enable what's called machine learning.

## Machine Learning Engineer

A machine learning engineer, or ML engineer, creates the software that we think of when we picture a computer learning. These programs have algorithms, or sets of rules or instructions to solve problems, that a computer must follow when presented with data. The software uses that data to make **predictions** and go a step further—it can learn.

### Fast Fact

Neural networks and deep learning are kinds of machine learning. A neural network is a method of computer processing based on the human brain and its **neurons**. Deep learning uses layers of neural networks.

**For AI to learn a user's preferences, it needs the user to provide data.**

For example, machine learning engineers work on artificial intelligence software for recommendation engines. Streaming services such as Netflix and Disney+ use recommendation engines, which use the data from a user's previous selections to suggest a movie or TV show. They predict the user will be interested in this entertainment. If the software is given the information that the user did not like the movie or show—perhaps through a rating system—it will recalculate future predictions with this new data and make a more **accurate** prediction of what the user will like next time. In this way, AI is considered to be learning.

**Music streaming services use recommendation engines to suggest songs that a user might like based on past data.**

Machine learning engineers often study computer science, **statistics**, or mathematics. For some workplaces, they may be required to have a master's degree—a higher degree they get after finishing college—in machine learning, neural networks, or deep learning.

## Natural Language Processing Engineers

Have you ever asked a virtual assistant what the weather forecast is or to play a certain song? If so, you can thank a natural language processing engineer. Natural language is how we speak (as opposed to computer code). Natural language processing (NLP) engineers focus on human language, both spoken and written. They help create software that can

# CHATBOTS AND VIRTUAL ASSISTANTS

If you use the internet, you've probably interacted with artificial intelligence. Chatbots are programs that use AI to "talk" to human beings, answering questions and finding requested information. The first chatbots were text-based. They wrote to computer users and collected data that was typed in return. Chatbot technology now uses spoken language too. It can be experienced through so-called virtual assistants, such as Amazon's Alexa, Apple's Siri, and Google Assistant. These understand voice commands, using them to return results to users. They use machine learning too. Machine learning engineers are key in building and maintaining these platforms.

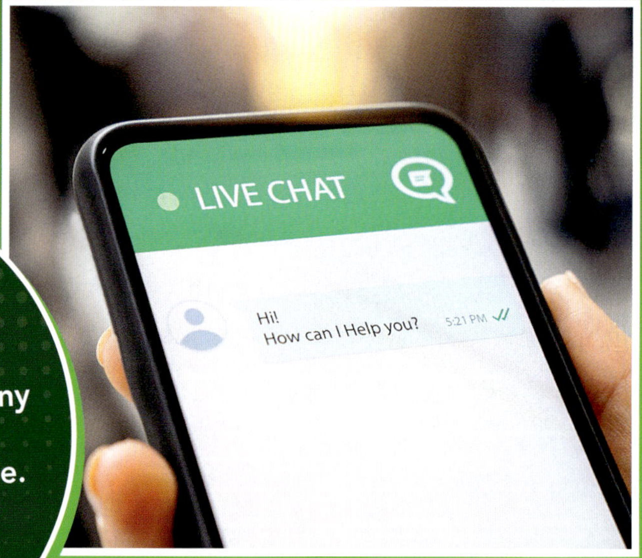


A chatbot often starts with a question. It uses words in the user's reply to calculate what information to provide the user.

**Fast Fact**

ChatGPT is a kind of chatbot that was created by the AI company OpenAI. It can make conversation with people. It can also be given certain tasks to do, like writing code.

**Many virtual assistants operate through smart speakers, which are connected to the internet and controlled by spoken commands.**

> **Fast Fact**
>
> An NLP engineer's job often involves separating sentences and pieces of text into bits of data that a computer can understand.

translate human text and speech into a language that a computer can understand, act on, and respond to, again using natural language. These engineers are especially important for virtual assistants, speech recognition programs, **dictation** software, and other systems that rely on human voice commands.

NLP engineers may seek a degree in computational linguistics. Computational means having to do with computers or calculations. Linguistics is the study of language. NLP engineers also study computer science, engineering, data science, mathematics, and statistics.

**Data scientists can create AI algorithms that help identify health problems in the human body.**

# AI AND INFORMATION

Chapter Three

Artificial intelligence is all about data: gathering it, examining it, and using it correctly. While you may think AI software just searches the internet for answers, it needs to be able to do this in a smart way. (After all, there are a lot of wrong answers on the internet!) For a computer to simulate intelligence, it needs to acquire the right data to reach its objectives. Some AI careers focus on these facts and figures.

## Data Scientists

Data scientists are important members of any AI team. They're experts in data, or information. They know how to collect it and study it so that they can see patterns that reveal something. Data scientists may use AI to help them gather data, or they may create algorithms that AI can use to make decisions based on data.

> **Fast Fact**
>
> Many data scientists work for businesses. They use AI to help identify trends, such as what people are buying and when.

Artificial intelligence needs data in order to become better at predicting and decision-making. Data scientists can filter down unstructured data—or large

**Data scientists are often needed to find the important parts in a set of data before that data is used for artificial intelligence.**

**Fast Fact**

One kind of data scientist is called a big data analyst. "Big data" is a term that means a large amount of information that's hard to manage. An analyst analyzes, or carefully studies, information.

collections of files such as emails, photos, and videos that aren't in an organized database form—into a usable and helpful package for AI.

For example, data science is used by some sports organizations to predict the impact athletes will have on a team. A lot of data is collected about sports, including speed, scoring, minutes played, games won or lost, and more. AI software wouldn't know what to do with all those numbers without a data scientist to help.

In school, future data scientists focus on many of the same fields of study as ML engineers: statistics, mathematics, and computer science. They need to be experienced programmers. They often need an advanced degree too.

## User Experience (UX) Designer

Have you ever visited a website or played a video game and had a hard time finding your way around? It probably wasn't designed well! AI is for people's use. Someone has to make the experience of using it a pleasant and easy one. That person is called a user experience (UX) designer. It's important that the user doesn't feel confused or overwhelmed when using a website or other AI environment. If they do feel that way, they might not want to use it again.

UX designers may do research to find out how people interact with **interfaces**. They may work with a group of users and ask them questions about their experiences. They may design a new interface

**A UX designer makes a product or service usable and enjoyable.**

## AI RESEARCHERS ASK, "WHAT IF?"

An AI researcher can be a creative thinker. Some are allowed to let their imaginations go wild as they figure out how to solve problems. Some work at a research institute or university. They may have more freedom about which real-world problems to try to solve. Others may work for a business trying to find a solution to a specific problem using AI. These scientists still need to know about coding, data analysis, and the many other parts of the AI-building process to know if their ideas can be made into a real-world application. AI researchers are needed for the next big breakthrough in AI.

AI researchers may see a need and think of ways to address it with AI-assisted technology, such as this hospital robot.

A UX designer wants to make the user's experience as intuitive as possible.

based on their research and collaborate with engineers and scientists to make it happen. UX designers may have a background in graphic design, software development, industrial design, or another field. They need experience but not necessarily an advanced degree.

## Fast Fact

In UX design, the outline of a website or **app** is called a wireframe. If that is approved, the designer makes a working model called a prototype.

Robots can take over many tasks that people do. They're especially used for repetitive jobs.

# AI AND ROBOTICS

Chapter Four

When people hear the term "artificial intelligence," they often think of robots. However, not all robots use AI. For example, robotic arms in manufacturing may be programmed to do one or two actions, such as painting or welding parts. These robots are not learning and changing actions as a robot with AI programming does.

However, the field of robotics is increasingly using AI to advance. Careers in robotics are an exciting way to shape the future.

## Robotics Engineer

Robotics engineers may oversee the construction of a robot from idea to completion or focus on one part of the process. They may also maintain the robot after construction. Robotics combines knowledge in a few fields, including mechanical engineering, electrical engineering, and computer systems. A robotics engineer may specialize in one of these, but should have some experience in all. In addition, those interested in AI need to know about machine learning in

**Fast Fact**

**In 2020, the World Economic Forum estimated that by 2025 robotics and other technology will have created 12 million more jobs for people than they have taken away from people.**

**Robotics engineers often use—and must therefore be skilled in using—computer-aided design (CAD) and computer-aided manufacturing (CAM) software.**

### Fast Fact

Hydraulics has to do with using liquid pressure to make a part of a machine move. Pneumatics deals with using compressed gas to make machinery move.

order to include AI software and make that software work with the robotic parts. Robotics engineers may work with machine learning engineers, natural language processing engineers, software developers, and others as well.

Robotics engineers may study mechanical, manufacturing, electrical, or computer science engineering in college. Their studies may include courses on microprocessors, hydraulics and pneumatics, electronic circuits, and computer programming.

This robotic arm uses pneumatics to create mechanical motion and pick up a box.

## Mobile Robotics Engineer

While many robots are fixed in place as they do their jobs, others need to travel. Some robotics engineers specialize in how the robots can travel and move around most effectively. This career path is called mobile robotics engineering.

A mobile robotics engineer has to make sure a robot can get around, whether on legs or wheels. Robots that are sent

This robot is delivering food in Los Angeles, California.

### Fast Fact

Autonomous vehicles are those that can operate without human control. AI is a necessary technology to make these self-driving cars safer for passengers.

# PERSEVERANCE

NASA's *Perseverance* is the newest rover, or exploration vehicle, on the surface of Mars and a great example of an AI success. About the size of a car, it's autonomous and can drive around the planet's challenging surface. It's equipped with AI software called Autonomous Exploration for Gathering Increased Science (AEGIS). It uses AEGIS to select rocks to study and collect data from—without input from NASA scientists and engineers back on Earth. *Perseverance*'s mission is to search for signs of life and gather rock samples that may be brought back to Earth at some point.

The *Perseverance* rover was built and is overseen by NASA's Jet Propulsion Laboratory in California.

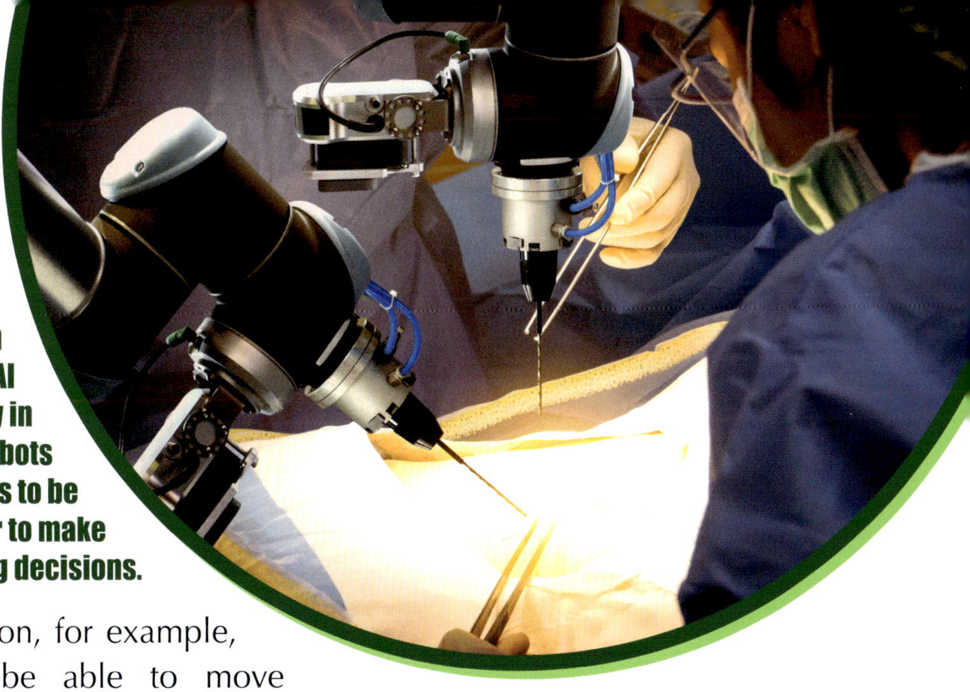

Many things can go wrong in surgery. AI technology in surgical robots needs to be superior to make lifesaving decisions.

to the moon, for example, have to be able to move around the surface. They also have to be programmed to make decisions as they explore, such as avoiding deep craters. Since the engineer cannot have frequent contact with a robot like this, AI is especially important.

## Future AI Experts

AI experts have their choice of industries to get involved in—from space to sports! People with the science, engineering, programming, and research careers explained in this book started with a STEM interest. You can start your career path today by studying especially hard in math and science classes. You should also look for opportunities to learn how to code. Some websites, such as code.org, offer free instruction.

Finally, you should keep up to date about AI issues. For example, some scientists are worried that AI could become so advanced that humans would be unable to control its actions. They worry about what that could mean for humankind. You could help find solutions to these and other concerns as a future AI expert.

# MORE USES OF AI

## Space

- rover and spacecraft operations
- mapping stars
- positioning of **satellites**

## Military

- unmanned aerial vehicles (drones) for monitoring dangerous sites
- analyzing social media for threats
- military training software

## Education

- grading material
- detecting cheating
- virtual learning assistants

## Mining

- autonomous trucks and drills keep miners out of unsafe conditions
- machinery to locate ore and sort wanted matter from unwanted matter

## Health Care

- illness and disease detection
- analyzing medical records
- personal medical devices

## Agriculture

- crop-monitoring drones
- robotics for planting, watering, fertilizing, and harvesting

# THINK ABOUT IT!

1. Why do you think many people are fearful of artificial intelligence?
2. Where else could AI robots be useful?
3. Do you think artificial intelligence such as ChatGPT should be allowed in schools and for schoolwork? Why or why not?
4. In what ways is artificial intelligence similar to and different from the human brain?

# GLOSSARY

**accurate:** Free from mistakes.

**app:** Short for "application." A computer or smartphone program that is made to perform certain functions, or tasks.

**collaboration:** Working with another person or a group in order to accomplish a task.

**complex:** Describing something with many parts that work together.

**dictation:** The act of saying words aloud so that a machine can record them.

**interface:** A program that controls how information is shown to a computer user and how the user can communicate with the computer.

**intuitive:** Easily understood or learned.

**neuron:** A cell that carries messages between the brain and other parts of the body.

**prediction:** A guess about what will happen in the future based on facts or knowledge.

**repetitive:** Happening again and again.

**satellite:** An object that circles Earth to collect and send information or aid in communication.

**statistics:** A subject related to collecting and studying numbers to show information about a topic.

**technology:** A method that uses science to solve problems and the tools used to solve those problems.

# FIND OUT MORE

## Books

Kulz, George Anthony. *Artificial Intelligence in the Real World.* Lake Elmo, MN: Focus Readers, 2020.

Lane, Dale. *Machine Learning for Kids: A Project-Based Introduction to Artificial Intelligence*. San Francisco, CA: No Starch Press, Inc., 2021.

Rathburn, Betsy. *Artificial Intelligence*. Minneapolis, MN: Bellwether Media, Inc., 2021.

## Websites

**Artificial Intelligence**
*kids.britannica.com/kids/article/artificial-intelligence/390648*
Read more about the concept of AI.

**Mars Mission 2020 Perseverance Rover**
*mars.nasa.gov/mars2020/*
Follow the rover as it demonstrates how AI can help humankind's search for knowledge about Mars.

**Will Artificial Intelligence Replace Human Jobs?**
*www.wonderopolis.org/wonder/Will-Artificial-Intelligence-Replace-Human-Jobs*
Find out more information about this pressing question of modern times on this website.

# INDEX